Digital Distress; Growing Up Online

COMPANION WORKBOOK

Dana M. Chandler

and

Lisa K. Strohman, JD, PhD

Digital Distress; Growing Up Online, Companion Workbook

By Dana M. Chandler and Lisa K. Strohman, JD, PhD

Publishing Services by KD Resources
Clarkston, Washington
kdresources.org

Registered with the United States Copyright Office

Library of Congress ISBN 978-0-578-27706-6

"In my experience as a principal and teacher, nothing gets a teacher more fired up then someone coming in, claiming to be an expert, but has never been in a classroom or tried the strategies they are pushing with students in the classroom." -Such an important statement! Teachers are in the trenches every day and try new theories and pedagogical practice for the benefit of their students. I think this workbook is an essential resource to accompany Dr. Lisa Strohman's text, Digital Distress; Growing Up Online. It requires educators, (teacher/administration, etc.), to think about technology and identify how much is too much. This workbook can be completed by a single educator individually or as a grade level team. In addition, school administrators could use this as a professional learning and book study for their staff in helping to better understand students, their technology usage, and how to balance this "addition" while in school.

One part of the text that resonated with me was: "It's imperative that educators understand the specific characteristics of this generation and adapt the classroom in order to reach them." This is very important! It is critical that we work to understand our kids and how we can adapt our classrooms to best meet the needs of our individual students. I also like that it addressed Gen Z and gender equities.

Generally speaking, there is a significant gap in students who use technology and the teachers wo are teaching these students. Educators have to understand the technology our students engage in and how to best meet their individual needs. We have to engage in reflective practice and ask ourselves, are we contributing to the problem or using technology as a means to engage students in collaborative 21st Century style learning? How do we support the balance our students need? Educators, schools, and districts should all be using this book!

~ Dr. Tonya Laliberte, Community Superintendent,

Dept. of Defense Education Activity

Contents

FOREWARD

FOREWARD

Let me introduce you to, Dr. Lisa Strohman, clinical psychologist, attorney, author, child advocate, and founder of the Digital Citizen Academy. She has devoted her life's work to educating others about the benefits and dangers of online use and helping others create a balance between their devices and the real world.

I ran into "Dr. Lisa" recently at my aunt's birthday party in Sacramento, California. I met her many years ago and our lives have intertwined throughout the years, but I had never had the opportunity to spend time getting to know her. During our conversations in Sacramento, I learned that Dr. Lisa had written books concerning children and their parents who were tackling the difficult task of balancing computer devices with the real world. We discussed her own children and how they have staggered their way through the challenges with no specific guidelines on how to create a balance in the time spent on devices. With her knowledge and experience, their family had more guidance than most.

I shared about my experiences with students in the classroom, particularly when the learning moved online due to the COVID quarantine that we had experienced. Our conversations led me to dive deeper into my own research once I returned home, as well as get the most recent of Dr. Lisa's books.

Reading, *Digital Distress; Growing Up Online,* was mind blowing! Dr. Lisa was able to say everything that I'd been contemplating and she and Dr. Westendorf had done the research to back it up. The book gave me the "why" to my questions about the change in students' social/emotional skills and behaviors and included great strategies for parents.

What the book was missing were those same great strategies for teachers. I called Dr. Lisa and we discussed the need. Dr. Lisa already had been working with school districts, traveling around the country to try to get this much needed information to those in the trenches with children, but this had been limited to what her personal schedule could build in. She, after all, was only one person and could only go to so many places. Her efforts have been gallant as she has been on news broadcasts, talk shows, and podcasts, as well as key noting at conferences, school districts and statewide educational trainings, trying to spread this much needed information to classrooms. She had a website built (*drlisastrohman.com*), started a blog, did some TEDx talks, and created her own podcast, called, *The Point*, in order to get her message out.

I offered to help, armed with a new-found motivation to get the word out. But first, I wanted to make sure that whatever information and classroom strategies that I would share were relevant and worked. In my experience as a principal and teacher, nothing gets a teacher more fired up then someone coming in, claiming to be an expert, but has never been in a classroom or tried the strategies they are pushing with students in the classroom. If I was going to share the information, it had to be tested at school.

My Parkway team came to my rescue. Parkway Elementary is a P-6 elementary school located in the heart of downtown Clarkston, Washington. The staff has changed significantly over the last few years, bringing in new faces and new ideas. I am included among the newbies, as I had only moved back home and been at Parkway for three years. (My first year was the year of the COVID 19 shutdowns.)

Parkway is typical of a modern school in eastern Washington State. It has free/reduced lunches for low income, but is blended with students from all social and economic backgrounds. Though there is not a large amount of racial diversity, Parkway's population is greatly diverse in family income. From students who live in cars or couch surf, to those who live in luxury and have all the devices on the market, the school attempts to provide as much equity as possible. Overall, the staff members are committed to this and work as a team to make sure all students have what they need during the school day.

Parkway is full of committed and dedicated teachers, as well as other staff, who work well together. They are models for respecting each other and supporting each other. Though we certainly had our conflicts, the leadership of Dan Lejameyer, kept us moving forward and providing the best instruction we could for kids. With this climate as a springboard, we dove into the book, *Digital Distress; Growing Up Online*.

The following pages are the information, ideas, reflections, and practical classroom strategies that we created as a collaboration. This book is not intended to be the only solution to teaching kids about digital citizenship and helping them to gain social/emotional awareness and growth. It is meant for educators to glean ideas and applications that will work for them. If you are an educator, there will certainly be ideas within this text that you will not use. However, I'm sure you will find some that are helpful and practical for your use in the classroom.

PREFACE

PREFACE

Digital Distress; Growing Up Online by Dr. Lisa Strohman and Dr. Melissa Westendorf is an exceptional book full of scientific research, personal notes, first hand stories, and suggestions for parental actions concerning children who have only known a world with the internet at their fingertips. The text was written with parents in mind. Sprinkled throughout the text are rich and actionable ideas for parents as they do the undaunting task of raising their children in a time that is unprecedented concerning the risks and dangers of the internet.

What the text does not include are ideas and suggestions for teachers in the classroom concerning how to engage, manage, and instruct children who have lived completely different experiences than the teachers before them. With the guidance of Dr. Lisa Strohman, the following pages are designed to give teachers the background knowledge, information, and self-reflection activities needed to be more effective in the modern classroom.

INTRODUCTION

INTRODUCTION

My hope is, that after reading the text, *Digital Distress; Growing Up online*, and going through the activities in this book, you will be more effective at reaching kids in the modern school.

What this book is? It's a collaboration of educators. We've put together meaningful activities that make teachers more effective. What it isn't? It is not all exclusive. We don't claim to have the answers for every problem in the classroom. It's also not designed as a curriculum. It is designed to be a self-study and teaching tool.

To get the most out of this self-study, I recommend doing it in partnerships or groups of teachers. However, it can be a good learning tool, if done alone as well.

Learning Model

This study is divided into four sections based on the book, *Digital Distress; Growing Up Online*, by Dr. Strohman and Dr. Westendorf. Each section includes a cluster of chapters from the book that all relate to the same learning model. It is divided as follows:

Section One: This includes the background and relative information regarding technology in the 21st century.

Section Two: This includes the areas of technology overuse and what that looks like.

Section Three: This includes the challenges facing us concerning technology use today.

Section Four: This section is full of hopeful solutions and teaching strategies for the classroom when facing those challenges.

Reading the book, *Digital Distress; Growing Up Online*, and completing the Companion Workbook, should take approximately eight to ten hours, depending upon whether you choose to read the text as a share-out or each read the entire text individually, as well as the amount of group discussions. Just remember that you will get out of this self-study what you put into it. The deeper you dive into the material, the more you will understand about students' use of technology and how to help them become healthier with their balance socially and emotionally in the classroom.

Learning Objectives

By reading and participating in this study, you will learn about the historical impacts of the internet surge in our world, the statistics for its use and overuse, what that looks like in our modern culture, the challenges we face from it, and what we can do about it at school. Overall, you will be learning about the background, research, social/emotional impacts, inequities, teaching technology citizenship, and how all of this might impact today's modern classrooms. It is the goal to help educators be more effective at teaching students to build stronger face-to-face human connections, while also understanding the appropriate use of virtual connections in their lives.

My hope is, that after reading the text, *Digital Distress; Growing Up Online*, and going through the activities in this self-study book, educators will be more effective at reaching kids in the modern school.

PRE-COURSE SURVEY

Pre-Course Survey-Answer the following questions before you begin this study.

How much tech is too much? What do you define as a balance?

How can student online use create problems at school?

Concerning cyber abuses, such as bullying or sexting, how do you respond to those who say, "It can't happen here. We're too nice at my school." Or "Our kids wouldn't do that."

What recommendations might you have to get students to be accountable for their own boundaries on time and content while online?

How do you think technology creates inequities among students?

Knowing that we can't control what happens to students at their own homes, how can we create a safe online culture in our classroom?

SECTION ONE - Background

SECTION ONE - Background

Instructions: Read Pages 1-24 of *Digital Distress; Growing Up Online*

Technology Today

I have had to fight a battle with children not keeping to an appropriate bedtime, angry for having to come to school, longing for the video game they were winning or having a tantrum for losing all their points or characters they so diligently created. The fits I have had to endure…and the excuses from the parents that rationalize the behavior. It left me shaking my head knowing it was a battle with the children and the parents. And to be fair…this is not ALL parents.

~Eleanore Follett, Third Grade Teacher

My parents were both teachers. I remember listening to them as they progressed through their careers reflecting on the students they were teaching. I heard phrases much like I hear today from teachers; "These kids today don't understand respect." Or "This generation of students haven't learned how to behave." I hear those same phrases today from educators and parents.

In the text, the authors discuss the label of "addict" and explain why they prefer to use the word "overuse." They define technology "overuse" to refer to children who "feel the need to use a piece of technology and [find] separation from that technology to be extremely overwhelming, resulting in rage, anxiety, depression, or manipulative behavior." According to the Digital Citizen Academy statistical information, 66% of parents felt their teens spent too much time on their devices, and about half of the students at the time agreed.

Digital Distress; Growing Up Online, defines the overuse of technology in children as something that creates behaviors that cause problems such as drug use, over eating, gambling, or playing video games over and over, and continuing that behavior despite negative consequences. In order to help students curb that behavior, we must recognize the negative aspects of the overuse of technology and find strategies to prevent it.

The study of the effects of the overuse of technology is in its infancy. However, recently, the World Health Organization has included it in the 2018 edition of the "International Classification of Diseases" language about Internet Gaming Disorders. With this classification, there has been some backlash regarding the inclusion of technology addiction (overuse) as a medical diagnosis. Critics have several reasons for their arguments.

One: Some believe that because technology is not a drug, it isn't a medical condition. However, research shows us that repeating the same compulsive behaviors over and over again, can have the same effect on our body and brain as if we were taking a drug.

Two: Another argument is that children who overuse technology is rare. However, from 2014 to 2018 the rate of teens reporting that they were on a device almost constantly, nearly doubled from 24% of teens who say they use the internet almost constantly to 45%. This trend is rising every year.

Three: Another criticism is that some believe that technology overuse should not be considered a mental illness. This issue has more to do with the treatment of the overuse. If technology overuse is considered an addiction and mental illness then there are many more players and billions of dollars involved. More questions arise such as, "How do we treat this unique addiction?" "Who will pay for it?" Many of these questions have yet to be answered.

Four: Another belief is that technology overuse is not caused by the technology itself, but by those who use it. It can be reasoned that the overstimulation to the brain from technology (internet, programs, apps, etc.) overuse impacts the brain like a drug, so it's not the technology actually doing the harm. This argument is similar to the chicken and the egg debate. Either could be true.

Five: Another important misunderstanding is that technology is not a unique addiction. The argument being that anything can be addictive if used or done inappropriately enough to change behavior. While this is true, it's important to note that tech and gaming companies are now onboarding psychology specialists to create products that "addict" the brain. This should be concerning for parents and educators alike.

Six: There are some that believe that technology itself does not lead to suicide. It is true that actual devices, such as pads, desktops, or phones have not been linked to suicides. However, the use and participation of the apps, programs, and internet have been linked to increased rates of depression in children. Is it the technology that causes depression or the child's use of the apps, programs, or the internet that causes the depression? Again, this is another chicken vs egg debate.

Which one or two of these arguments do you relate to the most? If so, how and why?

Instructions: ***As a team of educators discuss, "The Good, The Bad, and The Ugly" chart on page 14 of Digital Distress; Growing Up Online.***

DIGITAL DISTRESS

TECHNOLOGY IN THE 21ST CENTURY... A SAMPLING

The Good	The Bad	The Ugly
Always connected	No boundaries between personal and professional life	Companies collect thousands of data points on every American consumer
Connect with distant relatives, childhood friends/classmates, and former work colleagues	Online connections are a poor substitute for face to face interaction	Loss of friendships over ideological differences introduced online
Interact with companies, celebrities, and elected officials like never before	Professional repercussions for every interaction online	Outside agitators sow social discord with hate speech and appeals to racism, sexism,nationalism, xenophobia,etc.
Ability to buy goods from allover the world	Loss of local retailers	Worldwide sale of illegal drugs across the dark web
Virtual reality simulation scan transport people anywhere in the world	Virtual reality pornography undercuts romantic relationships	Livestreams of young people captured and sold on child pornography websites

Finding the Balance

It's really important that we find a balance between the time we spend on our technological devices forming virtual connections, and the time we don't when we can form genuine and authentic human connections.

The book, *Digital Distress; Growing Up Online,* was written in order to teach families how to find that balance. Technology does have its benefits and its uses, but it can also be abused and used in ways that can permanently cause developmental delays in a growing brain.

For the use of this self-study we will be using the term **Goldilocks Zone.** This term refers to that balanced place where there's not so much technology that we might form addictions, but not so little technology that we have a disadvantage concerning our access to learning. The Goldilocks Zone means that we have found that place where it's not too much or not too little, but just right.

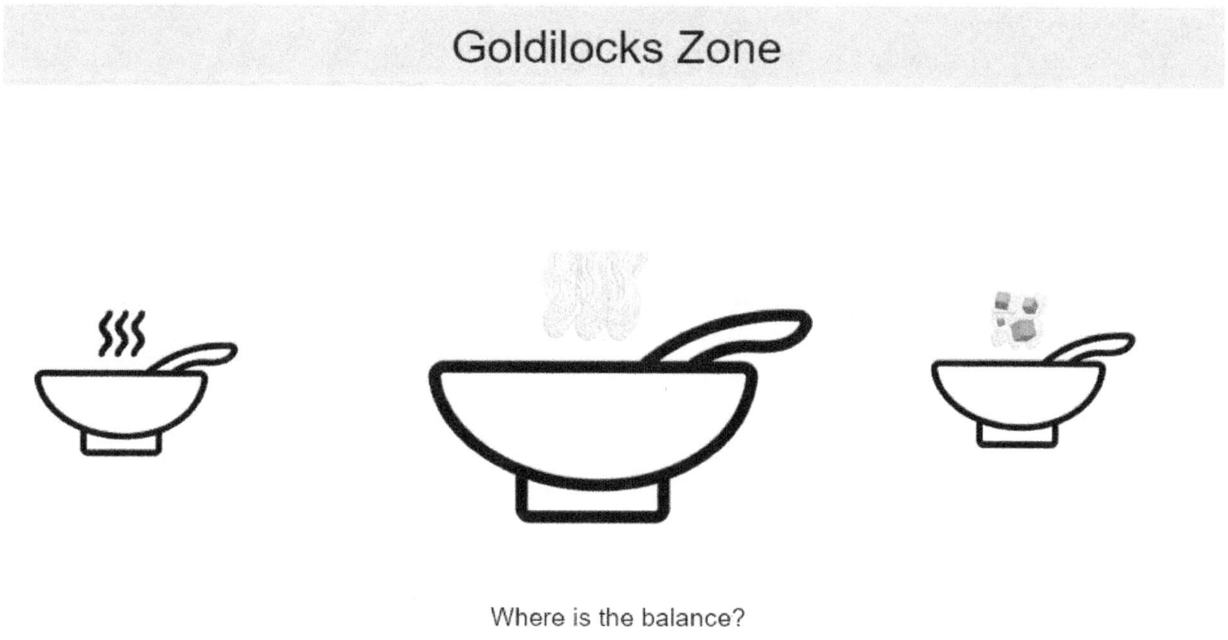

This illustration shows the concept of the Goldilocks Zone.

This healthy balance is so important as we grapple with the issues of social and emotional development in children. Finding that healthy balance in a classroom setting is what we will hope to achieve in the study. Our world is changing rapidly, and helping our students find a healthy balance with their devices will help them keep up with global changes without sacrificing their own development and learning.

Instructions: Read pages 25-36 of *Digital Distress; Growing Up Online*

Even though our underlying biology stays the same, social norms change, the economy changes, the culture changes, and the demographics of the country change, all creating an endlessly evolving stew of influences that make each generation distinct from that which come before and that which comes after.

~Digital Distress; Growing Up Online

Can it be that children in each generation simply are different and see the world from a new perspective based on the culture and social climate in which they grew up in? Research would tell us that's true. It is the old "nature vs nurture" question.

The most current research tells us that while we each come with a set of genetic instructions, those instructions can be modified by the environment that we live in and how we interact with that environment.

~Digital Distress; Growing Up Online

We know that the generation of students passing through the school doors view the world differently than the adults that are educating them. However, the changing cultural climate in which we live has developed a generation of kids with significantly different perspectives and view point from previous generations. It's imperative that educators understand the specific characteristics of this generation and adapt the classroom in order to reach them.

If you're like me, and in your 50s, you are further away from the characteristics of our current student population than most educators. While I truly believe age is not a disadvantage for being an effective teacher, it does create a larger gap between your early childhood experience, and that of the students in our modern classrooms. During this study, remember to keep an open mind and a growth mindset.

Defining Gen Z:

I'm trying to provide that "human connection" that is needed so badly. I've restructured my classroom so that we do most activities in cooperative groups. I use the cooperative groups for almost everything, from reading and writing assignments, and science tasks, to math problems. This one thing has increased my students' ability to solve problems. Now, if students are struggling with an argument from recess or earlier in the day, I can put them in a group and have them solve it among themselves, when before I would have had to intervene in order to settle things down enough to continue the learning. The cooperative groups have really helped my students communicate and understand each other in a deeper way.

~ Ashley White, Fifth Grade Teacher

During the discussion of this section concerning the characteristics of Gen Z (our modern students), the Parkway staff documented specific traits that we might not have thought of prior to reading the chapter in the book. We discussed how the experiences of our newest generation, plays an important part in how they see the world, and therefore, how they specifically need information presented in order to maximize their learning. Though our discussion topics were certainly generalizations and don't apply to every child, the characteristics that we noted as impacting a child's success in the classroom, are more prevalent than in previous generations.

The household structure of the past has changed, making most households either being managed by a single parent or both parents work full time jobs. This can make the "traditional" family life, where one parent is at home, less accessible. Parents are busy or overly preoccupied with just keeping the household in order. Spending time for school becomes a challenge. In addition, children in single parent households, more often, lack the financial advantages that two-parent households might be able to offer.

We made note of the fact that most kids in our classrooms don't naturally feel safe. They are bombarded with information about scary things in the world and internalize that into their everyday thinking. Issues like gun violence in schools or child abductions are not kept from them because it's all over the internet and in socially acceptable videos that children are exposed to. Because these issues are becoming more common or publicized more often, our childrens' sense of security and trust is malformed, much like a child during wartime.

Because of how scary the world can seem, this generation of children have higher rates of anxiety and struggle forming trusting friendships. They are living with their parents longer than any other generation before them. While they physically mature faster, they seem more immature and have a difficult time adjusting to being adults.

This generation of kids are the last white majority generation in the United States. They more easily tolerate differences and typically rally around the idea of our country's diverse population. They find it socially acceptable to openly discuss topics such as gender or race inequality.

However, with their increased exposure to video games and social media platforms the gender equities are widening. Boys are the typical users of video games and girls usually prefer social media for leisure times. As a result, boys are finding violence found in their video games more acceptable, while girls are trying to measure up to the unrealistic social and physical features that are portrayed on social media. Studies regarding these experiences with video games and social media correlate with significant increases in suicide ideations or successful suicide rates, especially in girls.

Today's children are natives of technology, while those of us in a different generation may feel like we are trying to speak a second language when we work with technology. This native vs immigrant concept effectively illustrates the gap of technology understanding between generations. Which is why, sometimes, we as teachers, feel like we are speaking a different language than our students.

But perhaps the most notable discussion that the Parkway staff had concerned parents. The book, *Digital Distress; Growing Up Online*, points out that parenting is very important in this whole process. While this has always been true, now more than ever, kids need a parent who is able and willing to learn the technology with their child, keeping them as safe as possible from online dangers. This can be a daunting task for parents! So how do we as educators support their efforts? While there are a few exceptions, most parents do care if their child succeeds at school.

What are some of the specific characteristics of Gen Z?

1.

2.

3.

4.

5.

6.

7.

8.

"Even though we may share the same biology, our realities are shaped by a different mix of experiences."

~Digital Distress; Growing Up Online

What is the difference between a digital immigrant and a digital native? Which one are you?

What are some of the gender differences within Gen Z? How might you provide better gender equity in the classroom?

How are the behaviors of students in Gen Z different than when you were a student at that age?

The Developing Brain

Instructions: Read pages 37-48 of *Digital Distress; Growing Up Online*

What do children need to promote brain development?

1.

2.

3.

4.

5.

6.

7.

8.

9.

10.

List the characteristics of a "Birth to Toddler" brain.

1.

2.

3.

Describe the brain during the birth to toddler years. Choose three of these needs and reflect on how you think it impacts the students in your classroom.

List the characteristics of a "Grade School" brain.

1.

2.

3.

What did you learn about the brain development of a child between about age 5 years old and adolescence?

List the characteristics of an "Adolescence" brain.

1.

2.

3.

Discuss two aspects of the adolescent years. How does that impact the students in your classroom? If they are in elementary school, what can you do to prepare them for adolescence? If they are in adolescence, what might you do to help them be aware of their development at this point?

Various mental powers peak at varying ages in our lifetime. What age group do you fall under at this time, and what do you think that means for you based on the information about the various mental powers and their peak on page 43?

On page 43 there is a chart that lists the age at which various mental powers peak. What is notable to you regarding this information?

On page 45-47, the authors discuss why parenting is so crucial for children, especially concerning their use of technology. How can you apply this information to the classroom? How can you, as educators, be a part of that modified environment students might need in order to meet their highest potential?

Reflect on what you've read about brain development in Chapter Three. Make a list of three things that you might try with your Gen Z students in the classroom in order to meet their brain development needs.

1.

2.

3.

Process Addictions and the Brain

Instructions: Read pages 49 - 58 of *Digital Distress; Growing Up Online*

As we continue this study, we must understand that our brain is flexible and adaptable. This can be a positive motivator, yet also a negative one. Every time we experience pleasure, our brain is rewarded by the chemical dopamine. While this experience is good in moderation, the release of too much dopamine is what happens during addiction. When there is too much dopamine released in an early developing brain, negative impacts can be identified. This typically looks like behavior problems, a lack of focus, and social/emotional delays, as the addict is continually trying to experience that "pleasure" response.

Unfortunately, in today's modern world, this practice of seeking the pleasure response from our brain is built into the technology products that we buy. Video games are designed to get people to constantly be seeking that positive reward in the brain, which makes the player continue the game and buy more of its products.

For gamers that spend more than three hours per day playing video games, the gray matter of the brain, which does most of the brain's processing and takes up 94% of the brain's oxygen, has been shown by research to shrink. The critical executive functioning of the brain and the empathy/compassion portions of the brain are the most impacted. In addition, the white matter of the brain, which is responsible for problem solving and multidimensional thinking, can also be compromised, slowing down processes, interrupting the communication connections within the brain, and short-circuiting nerve impulses.

Knowing this, what might you see from a child that is overusing video games?

What about a child who is overusing social media?

The most recent statistics put out by the Digital Citizen Academy shows that the decline in the appropriate use of technology is correlated with the introduction of social media platforms such as Instagram and Facebook. This has impacted the development of social and emotional skills in children.

According to the Digital Citizen Academy data, in 2021, the average teen spent about 14 hours per day online.

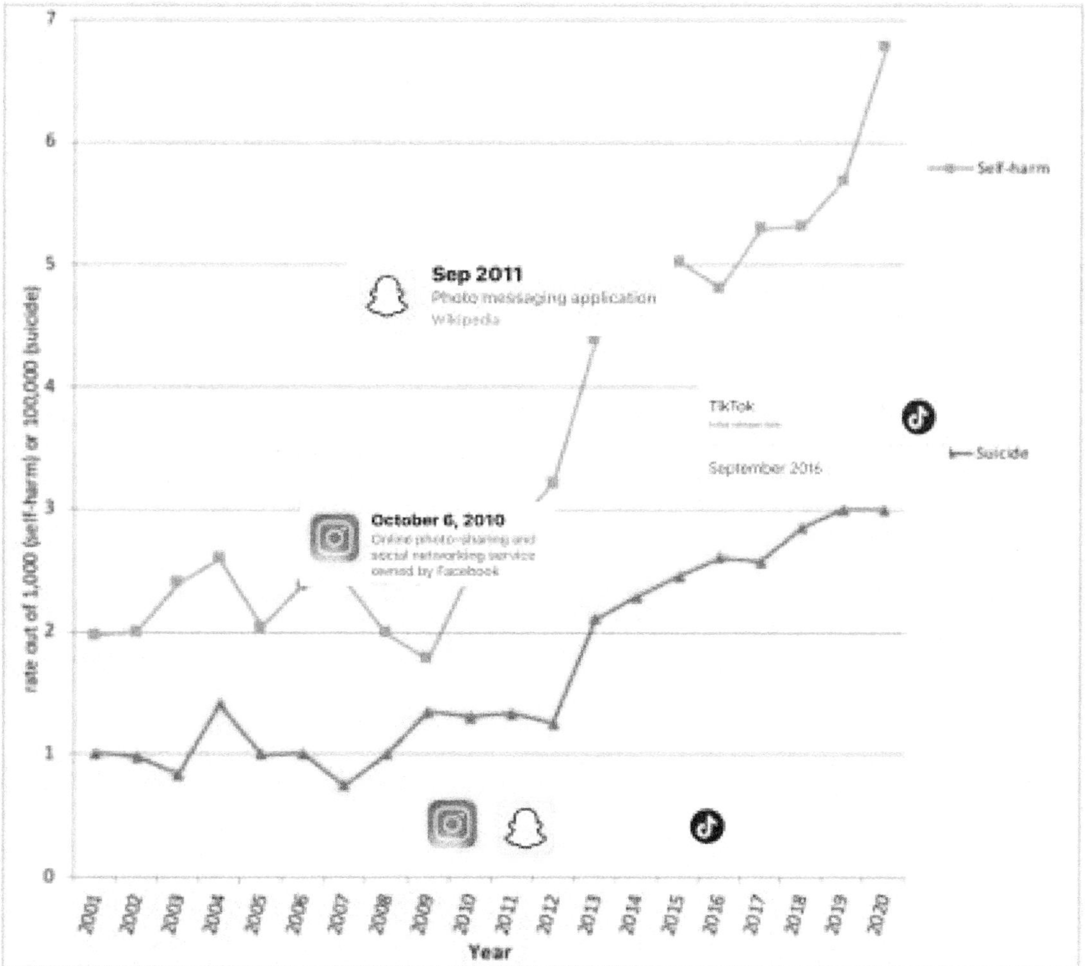

Figure 3: Emergency room admissions for self-harm and completed suicides, 12- to 14-year-old girls, 2001-2020. Source: CDC WISQARS database

Gaggle statistics, as the COVID 19 school closures were in effect, showed significant increases in student safety concerns.

The State of Our School Children
Gaggle Insight 2020-2021 School Year

SECTION TWO – The Four Dimensions of Technology Addiction

SECTION TWO-The Four Dimensions of Technology Addiction

Instructions: Read pages 61-133 of *Digital Distress; Growing Up Online*

The research put out by the Digital Citizen Academy shows that AWARENESS makes a huge difference in students' ability to monitor their own digital use.

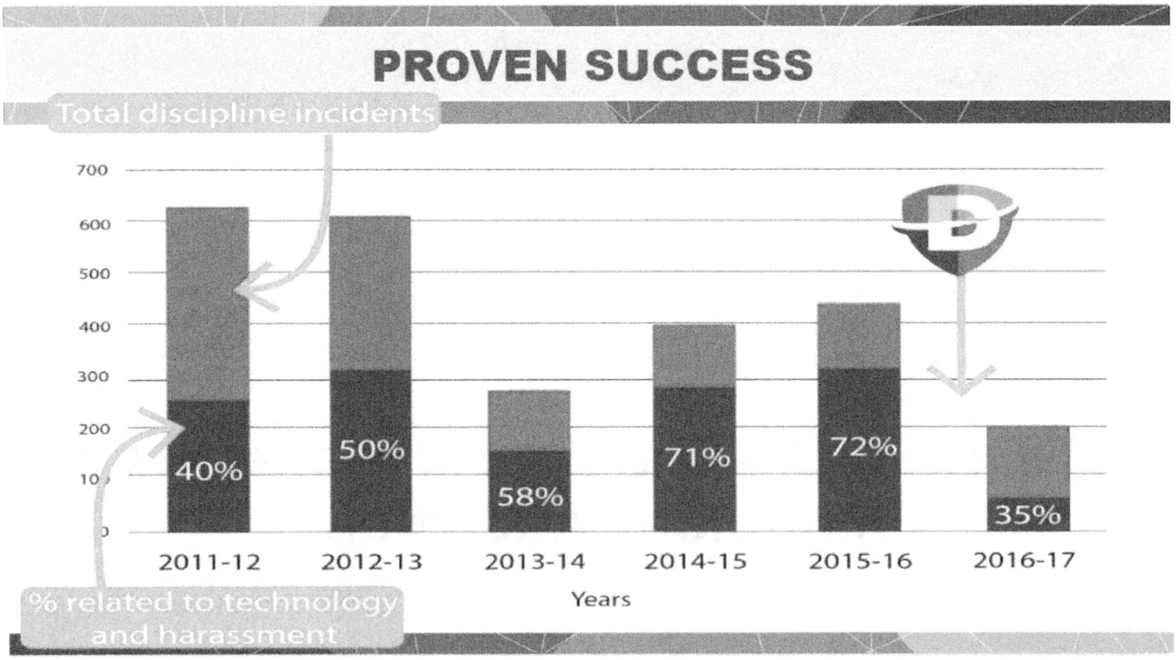

This chart shows the results of one middle school where a concerted effort was made by staff to train students on the risks of online behaviors and then compared the data using discipline referral related to harassment. You can see the difference in incidents significantly dropped immediately when the awareness campaign was implemented in the year 2016-17. This proves that most students are capable of successfully monitoring their own technology use.

Assessing Your Use of Technology in the Classroom.

Section Two of, *Digital Distress; Growing Up Online*, deals with identifying overuse or addiction to technology in children. The text points out symptoms of overuse and gives strategies to parents for managing the use of the devices at home.

In this section we will be dealing mostly with practical application of planning for technology use and making students more aware of how much time they are spending on their devices at school.

As educators, we can't diagnose an addiction or overuse problem. However, there are very obvious symptoms that we can note in the classroom, from behavioral, physical, emotional, and interpersonal symptoms. While parents have the responsibility of monitoring their child's online time during non-school hours, it's our responsibility to monitor that use during the school day.

To do that, we need to determine our own use of technology in the classroom and help our students be more accountable for themselves. This would not only increase our ability to create a healthy balance and technology use at school, but also help the students learn the life skill of balancing it for themselves when they are not at school.

What kinds of technology programs do you use in the classroom and what do you use them for?

____

____

____

____

About how many minutes during the school day are your students on a device of some kind?

If you aren't sure about the programs you personally use, what they are used for, and how long you use those programs and devices, the following Technology Log may help you get a more accurate picture of your classroom.

TECHNOLOGY USE LOG

Date:______________________________

Device	Time On/Off	Total Minutes	Purpose
ex: Chrome book	1.00-1:25	25	Study guide

Monitoring your own use of technology can give you a great example for your students. Once you've determined how much you use a device, it will be easier to guide your students through the same practice.

Assessing Your Students' Technology Use

Have your students each track one day of technology use in the classroom using the Technology Use Log below.

TECHNOLOGY USE LOG

Student Name:______________________________

Date:______________________________________

Device	**Time On/Off**	**Total Minutes**	**Purpose**
Chrome book	1.00-1:25	25	Study guide

(A copy of this log is located in the Appendix.)

Gather the logs for 1-5 days and see what your classroom results are. Is there one or two students on devices more than others? How did the students react to tracking their own time? What else did you discover?

Think of a student in your classroom who you are concerned has an unhealthy balance with technology. Complete this checklist concerning symptoms of the overuse of technology by that child.

Unhealthy Balance with Technology

◯ Constantly losing track of time while online

◯ Becomes agitated or angry when they are asked to get offline

◯ Would rather use the computer than socialize with peers or adults

◯ Has had discipline issues regarding their online use

◯Talks frequently about their online usage, such as game or social media posts

◯ Asks to be online rather than go outside to recess with peers

◯ Brings tech to school, such as phones, hand-held games, etc.

Do you have more students you could say have many of these symptoms? As you continue this self-study, hopefully, you will gain more ideas and strategies that can engage the students who overuse their devices.

Make a commitment to limit your technology use to only those times when it's needed. Talk with your colleagues about this. What ideas do you have to create a student contract or pact in your classroom in order to provide that healthy balance that students need?

What are the consequences for students who have technology discipline infractions at your school?

Do you think they are effective? Why or why not?

If you believe that the consequences at your school are not effective, do you have other ideas for how to handle technology infractions for the whole school, as well as the classroom setting?

Knowing that AWARENESS is the most effective strategy at getting students to self-monitor their own computer use, how do educators get parents involved?

How might the school, or you as an educator, make parents more aware of the issues concerning technology?

Think about the students you work with. How can students become involved in promoting a healthy technology lifestyle within the school? What kinds of activities can they do to increase awareness?

SECTION THREE - Facing Today's Challenges

SECTION THREE - Facing Today's Challenges

Before we delve into it, let's look at the most recent data from the Digital Citizen Academy on what the toughest topics are concerning students online.

In 2021, in the midst of the COVID 19 school closures, data indicated that children are online longer, increasing their exposure to cyber criminals and falling prey to predatory activity.

Not only are these children at higher risk at becoming victims of these online predators, their increased exposure and overuse of video games and social media make them more at risk for other addictive behaviors later in life, such as drug use, eating disorders, and more.

There are more than 750,000 predators online everyday

Children are on devices an average of 11 HOURS a day

There is a 97.5% INCREASE in predatory activity since covid

Instructions for Section Three of *Digital Distress; Growing Up Online:*

It's recommended that this section be read using four groups of educators so that each group can share the information that they've read, giving everyone access to the information simultaneously, allowing for professional discussions on each topic.

Recommendation for Educator Share Out:

Group One - Chapter 10; Video Games, Read pages 137-148

Group Two – Chapter 11; Pornography, Read pages 149-163

Group Three – Chapter 12; Social Media and Cyberbullying, Read pages 165-184

Group Four – Chapter 13; Children With Unique Needs, Read pages 185-201

(These sections can be read individually as well.)

The following section deals with very adult content, but is necessary in order to understand the very real risk to our children online.

Video Games/Gaming

After reading and discussing this chapter, what do you now know about the video gaming industry?

What are the problems with the overuse of video gaming and its impacts on kids?

How can video games help with learning in the classroom?

How are the genders portrayed in the video gaming industry?

What are the possible impacts on language from the use of home video games?

How has the use of home video games effected how children play?

Social media and video games are increasingly more vulgar, violent, (especially against girls and women), and sexual in nature. How might this impact the equity you see in your classroom?

Think of a girl in your classroom who seems to be overly impacted by the images portrayed on social media. What might you do as her teacher to help her?

Think of a boy in your school who seems to treat the girls as in a video game. What might you do as his teacher to help him learn more appropriate behavior?

Pornography

As you learn about pornography and its inclusion in the online traffic that children drive, let's get you up to date on the most recent statistics concerning pornography use.

According to the 2021 version of *Digital Distress; Growing Up Online*, pornography accounts for 35% of the internet content, with clicks out numbering Netflix, Amazon, or Twitter. 64% of young people from ages 13-24 look at porn once a week or more. Children under the age of 10 now account for 22% of the porn consumption online under the age of 18.

The website, InternetSafety101/EnoughIsEnough, states that 88% of internet pornography also portrays violence of some kind against women. And perhaps one of the most frightening statistics, is that those young adults who view X-rated materials are six times more likely to commit sexual assault.

According to the website webroot.com, that specializes in cyber security, 37 new pornographic videos are being made in the United States every day. Sex trafficking crimes are also on the rise.

The scientific conclusion is that viewing pornography as a child, increases that child's risk of having a sexual addiction or being a victim of a sexual crime.

DIGITAL CITIZEN ACADEMY

#MeToo / #TimesUp

$97 Billion Dollar Industry

Solicitation/Distribution - felony

Streaming Apps

Trafficking

What have you learned about the porn industry?

I think as educators, we can agree that frequent exposure to pornography has negative impacts on children. How does the viewing of pornography corrupt a child's brain?

What are the differences between the viewing of pornography by males compared to females?

How do you and your school prevent students from viewing pornography on a school device in class and out of class?

Social Media and Cyberbullying

In my 5th grade home room class we have been working through the social/emotional curriculum that our district uses. Our first month's focus was on the character trait of Respect. One of our lessons was focused on what we can do to be respectful online, as well as acknowledging ways that some people are not respectful. I was surprised by the level of awareness that this group of 5th graders had about all of the bad things involved in social media and online gaming. They were very aware of the negative things that can occur and were exceptionally open about sharing personal experiences of being involved in social media bullying. However, with all of this valuable information, they did not seem to feel any resistance against continuing to use certain platforms. TikTok and Snapchat seemed to be the most commonly used amongst this particular group and many of them heavily engage in posting and scrolling. Online gaming was another highly used platform with several reports of bullying from my students, yet they still engage in and play those same games. They see the negativity and experience it, but yet still desire that form of connection and status. What I took away from our class discussion was that despite the awareness of the dangers, they still habitually want these things in order to keep the human connections in their lives.

~ Raedawn Robinson, Fifth Grade Teacher

Mrs. Robinson's classroom is not unique. 43% of students report to have been bullied online and 1 in 4 claim that it has happened more than once. 70% report seeing frequent bullying online, and 68% agree that it's a real problem. However, only 10% of students will report an incident to a trusted adult.

While girls are twice as likely as boys to engage in bullying behavior or be a victim of it, 75% of all students report having visited a website bashing another student. And perhaps the most disturbing, is that bullying victims are 2 to 9 times more likely to consider suicide.

The statistics from 2021 are very concerning regarding children and self-harming behaviors. There have been dramatic increases in eating disorders, self-injury behaviors, such as cutting, and suicide rates among young people. In children ages 10-14, suicide ideations have skyrocketed 200% in girls and 50% in boys from pre-COVID years.

What is cyberbullying?

Who are easy victims of cyberbullying and why?

Knowing that 70% of students report that they have witnessed cyberbullying, how might we as educators teach students how to respond appropriately when they witness it?

What system is set up in your school for you to report when a student speaks about self-harming behavior or thoughts of suicide? What should you do and what will happen once you do report this?

Schools are limited in how they can handle cyberbullying situations because incidents don't necessarily happen at school or on school equipment. However, most schools do have disciplinary policies that allow for action if the cyberbullying is a "disruption" to the school or classroom.

Do you know your school's policy on cyber bullying? If not, how would you find out?

Social Media

As social media grows in advanced technology using filters and other physically enhancing video tools, especially girls are increasingly uncomfortable with their own bodies.

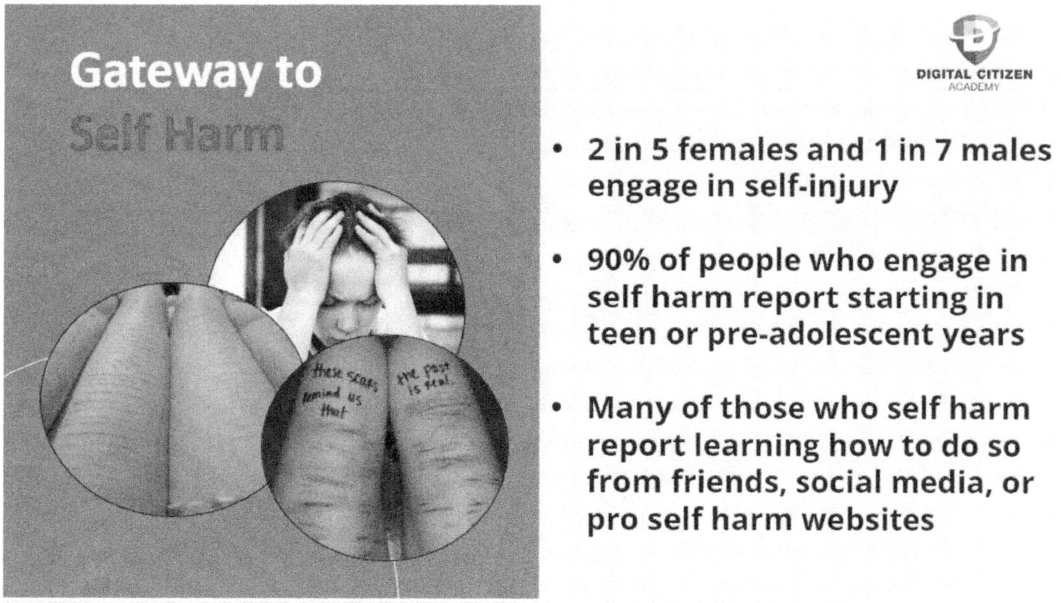

The trends of increased eating disorders and self-harming behaviors among girls is disturbing. How might the school as a whole, and you as an educator, be more effective at teaching girls self-worth and about their own value?

The trend of increased violence and aggressive vulgar language among children, especially boys, is also disturbing. How might you address this as a school, and as an educator in your own classroom?

Children With Unique Needs

Too much online exposure impacts some more than others. Students with neurosensory issues, such as ADHD (Attention Deficit/Hyperactivity Disorder) or Autism, are at a greater disadvantage then most students because of their ability, or lack thereof, to decipher information in the device format. They are at greater risk for technology overuse because their brain isn't always able to control the reward impulses given by the dopamine neurochemical. Transitioning from a device to another activity can be particularly difficult.

How would you manage a technology plan for a student with Autism, or another neurosensory disordered student, in your classroom?

What do you think an Autistic child would want you to know about planning for their technology use in your classroom?

What strategies might you teach an ADHD child concerning their own self-monitoring of technology use?

SECTION FOUR: There is Hope

SECTION FOUR; There is Hope

Instructions: Read pages 205-220 of *Digital Distress; Growing Up Online*

Mistakes Happen

Understanding that you don't have to be perfect in this process is part of the growth mindset that we hope you have and promote in your classroom.

Join with a colleague or two and discuss ways that you can assist each other in creating more effective technology plans in the classroom. What did you discuss?

____

____

____

____

____

How might you get students involved in the school and your classroom to create a more effective technology plan, giving a healthy balance to everyone during the school day?

____

____

____

____

Making students aware of the dangers of technology and its overuse is important. Research indicates that parents and educators should start when children are very young in teaching them to be a wise and balanced digital citizen.

Have you witnessed an example when a young child self-monitored their own technology use? If so, tell about that.

What this self-study shows is that those students in Gen Z lack specific social/emotional skills that prevent them from maximizing their learning.

Brainstorm a list of skills that the students you work with are lacking, such as *kindness*, *empathy*, or something as simple as *taking turns*.

At Parkway, we all agreed that making *human connections* was the most important skill that students were lacking, so we generated a list of ideas we could use to build the learning of that skill into our everyday routine. We then made a list of everything we thought might positively impact their growth in forming genuine trusting face-to-face connections.

Use the blank "Targeted Skill" template in the APPENDIX to create a plan for teaching these skills in a consistent intentional way. Write the targeted skill at the top and generate ideas and strategies for the daily/weekly routines in your classroom and specific activities that can be done to build this skill in students. It's important that the plans you use in your classroom is personalized for you and is easy for you to do.

SAMPLE

Digital Distress: Growing Up Online (Social/Emotional Skills)

Targeted Skill: Self-Respect

Rules/Routines/Frequency 1. Positive talk only "turn negative comments into a positive" statements and follow through 2. Practice problem-solving to help build trust/communication/respect for others and then to self-respect 3. Show examples of sad/angry etc. ... and examples of respect and kindness
Activity/Grade Level 1. Kids take turns looking in mirror and saying positive affirmations to themselves such as "I am smart." "I am brave." 2. Positive comment board – use post-its to put on board posted in the classroom and discuss 3. Glad Notes – Every student writes a positive comment for only one student per day.

As the Parkway educators worked, we came up with strategies that we could intentionally do in our classrooms to increase a variety of skills that Gen Z students were lacking. This list of these skills, along with targeted rules, routines, and activities that can be used in the classroom are in the Appendix.

I'm a lifelong gamer or, "digital native." I often tell my students that math is like a video game. Every video gamer can recall that level that feels insurmountable. You try and try and you just can't beat it. There comes that moment where you either give up or try a different strategy and persevere. After nearly wanting to throw your controller into the TV, you finally master that boss who previously seemed unbeatable. Athletes and gamers alike know and treasure that feeling: When you look at yourself and realize you did the impossible. There's nothing quite like it. And you can't manufacture that feeling. Switching the game onto easy mode won't bring the natural elation of a job well done.

~Charles Chambers, Sixth Grade Teacher

There are more examples of ways to address targeted skills in the APPENDIX based on the work that the Parkway staff did. On those templates, the last line is blank giving you the opportunity to add your own ideas to each targeted skill. You don't have to use any of the examples, of course, but we wanted to make sure that you had a bank of ideas to help you build a Goldilocks Zoned classroom.

REFLECTION

Reflect upon your learning. Now that you've read the book, *Digital Distress; Growing Up Online*, and have gone through the activities in this book, what are some of the things that stick in your mind as helpful, as you move forward in your classroom?

POST COURSE SURVEY

Post Course Survey- Answer the following questions before you complete this study.

How much tech is too much? What do you define as a balance?

How can student online use create problems at school?

Concerning cyber abuses, such as bullying or sexting, how do you respond to those who say, "It can't happen here. We're too nice at my school." Or "Our kids wouldn't do that."

What recommendations might you have to get students to be accountable for their own boundaries on time and content while online?

How do you think technology creates inequities among students?

Knowing that we can't control what happens to students at their own homes, how can we create a safe online culture in our classroom?

How have your answers changed from the Pre-Course survey you took at the beginning of this self-study?

Is there anything else you've learned worth noting, that you will want to remember later?

EPILOGUE

EPILOGUE

As you complete this self-study, keep in mind that there are specific social/emotional skills that will need to be intentionally introduced in your classroom to make students successful. The templates created in the Appendix have many of them listed, along with strategies for building rules and routines in your classroom and activities to teach those specific skills. Each skill has a space for you to add more to this resource should you so choose. Use these as you wish.

As you conclude this study, I'd like to leave you with the words of Mrs. Follett, a third grade teacher;

My class is the lowest I have ever experienced academically and socially in my 25 plus years of teaching. They struggle as readers. Math is a fear that must be conquered. And writing……what is that????
My hopes for this year:

- *To get the children acclimated to how to behave in school.*
- *To keep them engaged in their learning and take ownership.*
- *To realize an education is a privilege and not drudgery.*
- *Life exists beyond video games and that getting angry and throwing a tantrum isn't healthy for anyone.*
- *There are a number of potential friends to be made and kept for years to come.*
- *Working in a group or with a partner enhances learning and keeps the mind fresh with alternate ideas and ways of doing things.*
- *There is grass out there! There are games to play outside! The sun is shining; the sky is full of amazing cloud animals; the leaves are changing colors for reasons to be explored; and there are snowmen to be built and snowballs to be thrown!*
- *I want to teach them how to play kickball and I WANT to play!*

~Eleanore Follett, Third Grade Teacher

Mrs. Follett's hopes for the students in her class are mine for yours as well.

Sincerely,

Dana Chandler

ABOUT THE AUTHORS

Dana Chandler

Dana Chandler, grew up in a large family in the back country of Northern Idaho. Though raised in a small town, her parents instilled the value of civic duty and global responsibility.

Chandler graduated from Washington State University in Pullman, Washington, with a BA in Education/Special Education and then a MA in Administration, Curriculum, and Instruction from Gonzaga University in Spokane, Washington.

Chandler has experience as a special education and general education teacher, as well as being a school principal for a majority of her career. She has worked within the public school system and the Department of Defense Education Activity (overseas Japan) providing leadership, professional development and emergency response coordination.

Chandler has traveled around the world advocating for families and children as both an instructor for professionals, a teacher to parents, and keynote speaker at conferences.

As a former school principal and special education teacher, Chandler has seen the impacts of technology overuse in children. Her life's work has been to assist teachers in effectively teaching and caring for children. Chandler considers *Digital Distress; Growing Up Online, Teacher Edition*, part of a high-quality professional development program for educators assisting them to better understand and serve students in modern school classrooms.

From a family of writers, Chandler started writing at a young age. Finding a creative outlet in writing, Chandler, along with her identical twin sister, founded KD Resources in 2015, which provides professional development for educators, as well as coaching, and publishing services for beginning writers.

Currently, Chandler is an author, writing coach, and professional development specialist providing highly rated learning opportunities regarding effective teams, engaging classrooms, positive behavior supports, and social/emotional learning.

Chandler's other published books include *Lil*, *When the World Shakes; A Personal Account of the Great Tohoku Earthquake and Tsunami*, *Hello! I Have Autism*, and the *Backpacker Bob Reading Activities Guide for Raging Waters, Creature in the Woods, and Fire! Fire!*.

And, most importantly, Chandler is a Nana to three inquisitive grandchildren.

Dr. Lisa Strohman

Dr. Lisa Strohman has widely become known for her advocacy and education around mental wellness as it relates to our digital lives. She has worked with thousands of parents, schools, and children around the globe. An attorney, clinical psychologist, and author, Dr. Strohman established the Digital Citizen Academy, a non-profit program offered to schools with an in-home plan that educates, empowers, and inspires balance and prosocial use of technology.

She is the founder and director of Digital Citizen Academy, one of the first organizations to address the global issue of technology addiction and overuse. As a licensed clinical psychologist, Dr. Strohman has spent more than a decade working with individual, family, and adolescent clients struggling with issues including depression, anxiety, addiction, and technology overuse.

Recognizing the growing challenges parents now face with their children, and the increasingly serious issues her young patients are dealing with due to our growing dependency on technology, inspired Dr. Strohman to launch the Technology

Wellness Center, which she co-founded with Dr. Melissa Westendorf JD, PhD in 2014.

With her work in schools and in the media Dr. Strohman recognized the need for prevention education and diversion programs and founded Digital Citizen Academy for students, educators, and parents addressing these issues related to technology.

Dr. Strohman attended the University of California, Davis where she received a Bachelor of Science in Psychology, graduating Magna Cum Laude. She then completed a joint, integrated program in law and Psychology at Villanova and Drexel Universities.

During her undergraduate career, Dr. Strohman gained experience as a lead program coordinator with Families for Early Autism Treatment (FEAT), as a family reunification therapist and a hospice counselor. While attending graduate school, she performed research in substance abuse, risk communication and child abduction. Her avid interest in public policy earned her the honor of becoming a legislative intern for Congress. She was also selected as an honors intern with the Federal Bureau of Investigation and invited to become a visiting scholar with the FBI's National Center for the Analysis of Violent Crime division as she completed her dissertation.

After earning her Juris Doctorate, Strohman worked at a large law firm, while completing her residency at the Arizona State Hospital in both a clinical and forensic rotation in clinical psychology. She eventually settled back into her counseling practice as a clinical psychologist with Lifescape Medical Associates, where she utilizes her training in cognitive behavior therapy to help patients.

In addition to establishing the Technology Wellness Center, Dr. Strohman and Dr. Westendorf developed an assessment test to help parents identify if their child is at risk from technology overuse. The two have also co-authored two books; *Unplug: Raising Kids in a Technology Addicted World,* published in July 2015, and *Digital Distress; Growing Up Online,* published in 2021. Dr. Strohman has also written a children's book called, *Phone Alone*, published in 2022, using graphic art to engage children regarding the use of their devices.

In an effort to educate parents, caregivers and students, Dr. Strohman is a guest speaker to schools, parent organizations, and students to help raise awareness and offer preventative tools and resources. She is also frequently featured in the media as an expert source for issues related to technology use and behavior, including a weekly feature on Dr. Drew's radio show as well as an expert on Dr. Oz's ShareCare site.

THE DC ACADEMY

DIGITAL CITIZEN

The DIGITAL CITIZEN ACADEMY is dedicated to educating and training kids, school, and parents to become responsible digital citizens.

Founded by Dr. Lisa K Strohman, the mission of the DIGITAL CITIZEN ACADEMY is to educate schools, homes, and kids through our prevention and diversion programs, events, and resources. We focus on eliminating core issues like cyber-bullying, child luring, and suicide.

Programs offered by the DIGITAL CITIZEN ACADEMY give parents an opportunity to experience interactive lessons with their child. Home and school programs include module lessons that are grade specific, K-12, and include informative videos, activities, discussion questions, and online resources that make learning about digital safety and etiquette entertaining and fun.

To learn more please visit the website: digitalcitizenacademy.org

APPENDIX

TECHNOLOGY USE LOG

Student Name:______________________________

Date:______________________________________

Device	**Time On/Off**	**Total Minutes**	**Purpose**
Ex. Chrome book	*1.00-1:25*	*25*	*Study guide*

Digital Distress: Growing Up Online

Teaching Social/Emotional Skills

Targeted Skill: Human Connections

Rules/Routines/Frequency
1. Greeting upon entry to room: handshake/hug daily 2. Share out – Monday thru Friday 3. Turn and Talk 4. Collaborative Group Work - daily and/or weekly 5. Social/Emotional Training - Daily or Weekly (Second Step, Zones of Regulation, etc.) 6. Digital Citizenship Training for Students - weekly 7. Class Meetings 8. Tech-Free Zones or Times 9. Intentionally designated play groups at recess time 10. Structured games at recess 11. Role Playing 12.
Activity/Grade Level
1. Handshakes chosen by the class 2. Sharing important information about self 3. Weekend share outs 4. Class quotes 5. Share Board – Students answer question on the board 6.

Targeted Skill: Self-Respect

Rules/Routines/Frequency
1. Positive talk only "turn negative comments into a positive" statements and follow through 2. Practice problem-solving to help build trust/communication/respect for others and then to self-respect 3. Show examples of sad/angry etc. … and examples of respect and kindness 4.
Activity/Grade Level
1. Kids take turns looking in mirror and saying positive affirmations to themselves such as "I am smart." "I am brave." 2. Positive comment board – use post-its to put on board posted in the classroom and discuss 3. Glad Notes – Every student writes a positive comment for only one student per day. 4.

Targeted Skill: Kindness

Rules/Routines/Frequency
1. Weekly discussion about what kindness looks like 2. Daily share out about a way they or someone else has been kind 3. 3X per week watch Kindness 101; "On The Road With Steve Hartman" 4.
Activity/Grade Level
1. Group kindness games 2. Writing notes of kind words and put them on lockers or desks to give to other students 3.

Targeted Skill: Respecting Others

Rules/Routines/Frequency
1. Class Jobs 2. Taking Turns 3. Looking at someone using eye contact 4. When student is absent, put their name up saying, "We miss you." 5.
Activity/Grade Level
1. "I feel messages" 2. Talking Stems – (I agree/disagree with you because…) 3. How to respond to frustration- Discuss strategies as a class 4. Teach how to respect and react to differences, students with special needs, not spreading rumors 5.

Targeted Skill: Reducing Anxiety

Rules/Routines/Frequency
1. Partner works on a regular basis 2. Give students choices 3. Have peer mentors 4.
Activity/Grade Level
1. Use sentence starter templates 2. Use writing prompts 3.

Targeted Skill: Parents As Partners

Rules/Routines/Frequency
1. Send Newsletters – Weekly, Monthly or Quarterly 2. Phone calls home – Monthly or Quarterly 3. Daily check ins/outs 4. Use communication APP; i.e. Class Dojo 5. Send home volunteer notices regularly 6.
Activity/Grade Level
1. Students record their own behaviors and those records are either sent home with students or posted in their online classroom for parents to view 2. Classroom targeted parent information sessions-created for parents of specific kids who are struggling 3. Parent conferences for positives, as well as negative behavior or performance. 4.

Targeted Skill: Feeling Safe

Rules/Routines/Frequency
1. Greeting at the door with positive interaction for each student daily 2. Create a safe/quiet space within the classroom for students to go to in order to reset their behavior 3. Predictable routines and consistent rules and enforcement 4. Practice safety drills so students know what to do in advance 5.

Targeted Skill: Empathy

Rules/Routines/Frequency
1. Class discussions about empathy; "How would you feel if…" – daily and then weekly 2. Discuss examples of empathy – daily and then weekly 3.
Activity/Grade Level
1. Teach specifics of what empathy is and why it's an important characteristic. 2. Empathy BINGO 3. Find some books on Empahty and read them to students. Share real life stories of empathy from children; i.e. "On the Road with Steve Hartman" 4. Write letters to Veterans 5.

Targeted Skill: Sympathy

Rules/Routines/Frequency
1. SEL groups daily 2. Schedule post-recess drama discussions and create a plan to deal with it 3.
Activity/Grade Level
1. Second Step (K-2 at least) 2. Storyline Online related stories; "Lotus and Feather" 3. Send cards/pictures/texts to friends who have recently lost someone 4.

Targeted Skill: Phone Etiquette

Rules/Routines/Frequency
1. Phones put in backpack or basket in the classroom - daily 2. Outline consequences for breaking the phone rules – at least weekly 3. Reminder of phone rules daily, then weekly 4.
Activity/Grade Level
1. Group shares-each has a different topic (All grades) • How does if make you feel when someone is texting while you are talking with them? • What should your phone be doing while you're eating a meal? • How do you use social media? 2. Read Phone Alone using the doc cam– Have a class discussion (All grades) 3. Create phone rules as a class (Upper grades) 4.

Targeted Skill: Cyberbullying and Hate Speech

Rules/Routines/Frequency
1. Class Meetings 2. Consistent classroom rules clearly stating the consequences 3.
Activity/Grade Level
1. Lessons designed to define Cyberbullying and hate speech and why it's not appropriate 2. Discuss current events concerning an example of cyberbullying or hate speech (upper grade levels) or a school example (lower grade levels) 3. Lessons discussing the risks for both the victim and the perpetrator 4.

Targeted Skill: Preventing Hopelessness

Rules/Routines/Frequency
1. Class Meetings 2. Regular Peer Partnerships 3. Regular chants by the class repeating daily positive mantras as a group 4.
Activity/Grade Level
1. Friendship or encouragement groups at recess 2. Intentional partnering with students 3. Peer coaching - Use an older student to build a connection with a younger student 4.

Targeted Skill: Truth vs False

Rules/Routines/Frequency
1. Create a positive dependable classroom culture 2. Consistent and fair rules and routines 3.
Activity/Grade Level
1. Playing Two Truths and a Lie – Teacher gives first example 2. Discuss what are credible sources to get information 3. Students research topics to discern fact vs fiction 4.

Targeted Skill: Student Accountability/Responsibility

Rules/Routines/Frequency
1. Daily jobs 2. Routines where students are responsible for their own materials; both set up and putting them away 3. Classroom rules review clearly defining the expectation and the consequences – daily then weekly 4. Routines where students are responsible for turning in or charting their own work or progress 5.
Activity/Grade Level
1. Students write their own progress report. Have them give reasons why they have the score/grade that they do. 2. Student assignment books 3. Provide students time (and assistance if needed) to self-organize 4.

Targeted Skill: Off the Grid

Rules/Routines/Frequency
1. Scheduled "No Tech" Times or areas within the classroom 2.
Activity/Grade Level
1. Playing board games, blocks, puzzles 2. Creating word finds or crossword puzzles using vocabulary without tech help-Students can then have another student solve their work 3. Art activities; paint, crafts, coloring, etc. 4.

Targeted Skill: Preventing Aggression/Violence

Rules/Routines/Frequency
1. Class Meetings – How do we handle frustration or anger 2. Use Zones of Regulation to help students describe feelings daily 3. Class rules that are specific and clear about aggression 4. Recess groups with counselor for students who struggle 5. Set up a safe/quiet space within the classroom for the student to calm down 6.
Activity/Grade Level
1. Teach the Zones of Regulation – Put a poster up in the classroom 2. Teach strategies for dealing with anger appropriately 3.

Targeted Skill: Compassion

Rules/Routines/Frequency
1. Class discussion on what compassion is and why it's important 2.
Activity/Grade Level
1. Have a canned food drive at the holidays 2. Write letters to Veterans 3. View a segment about someone being compassionate in the "On the Road With Steve Hartman" series. Discuss it as a class. 4.

Targeted Skill: Manners (Appropriate Language Interactions)

Rules/Routines/Frequency
1. Expect the use of "please" and "thank you" consistently - daily 2. Class meetings having students use proper greetings, please/thankyous, turn taking, polite listening, etc. 3.
Activity/Grade Level
1. Have students practice greeting each other; say "hello" with eye contact 2. 3.

Targeted Skill: Gender Stereotypes/Equity (Age Appropriate)

Rules/Routines/Frequency
1. Class Meetings – Boys and girls can all do the same things (K-5 or 6) 2. Encourage all children to share their feelings with you 3. Address a child's negative stereotyping immediately so they don't gain a false stereotype of another child in the classroom. 4.
Activity/Grade Level
1. Class discussion about jobs in the community – Can girls and boys do them? (lower grades) 2. Use books and videos in the classroom that show a wide range of characters of diverse races and learning needs. Balance them between boy/girl main characters. 3. Teach respect for differences. 4.

Targeted Skill: Preventing Carelessness/Recklessness

Rules/Routines/Frequency
1. Set rules requiring students to redo work that is carelessly done 2. Class meeting; discuss why quality work and actions matter 3.
Activity/Grade Level
1. Teach students to carefully look at assignments; i.e. read them, underline important information, underline important words 2.

Digital Distress: Growing Up Online

Social/Emotional Skills

Targeted Skill:

Rules/Routines/Frequency 1.   2.   3.   4.
Activity/Grade Level 1.   2.   3.   4.

DIGITAL CITIZEN ACADEMY

SAMPLE ACTIVITIES

DIGITAL CITIZEN
ACADEMY

Kindergarten Activity

Name: ______________________

DIRECTIONS:
Draw a picture showing how you can be kind to others. Write a sentence about how you can be kind.

DIGITAL CITIZEN ACADEMY

Grade 2 Activity

Name: ______________________

WHAT WOULD YOU DO?

DIRECTIONS:
Take turns reading the cards and explain how you would relate to the situation as a Digital Citizen.

You are playing a game on a website and someone that you are playing with asks for your address.	A friend tells you that someone was being mean to them online.
You are researching on the computer for a project when you come across a bad website.	A friend gives you their password to log onto a website.
Your friend took an embarrassing picture of a classmate and told you that they are going to post it online.	You snuck your tablet to school and took it to recess. When you get home your mom asks why your tablet isn't in your room.
Everyone is making videos and posting them online. Your parents told you not to. When you go to a sleepover, a friend tells you to do it and that your parents won't find out.	You see a classmate playing a computer game instead of working on an online assignment. The teacher is working with a group and doesn't know.
You get a new video game and you really want to stay up late playing it. It's a school night. Your parents are asleep and won't find out.	You are working on a computer project with a group. Everyone is arguing about what pictures to put in the slideshow.

DIGITAL CITIZEN
ACADEMY

Grade 3 Activity

Name: ____________________

Roll a Story

1. Partner one rolls dice.
2. Look at the chart for the number rolled. Make up a story using the character, setting, and problem on the line that matches your roll.
3. Make sure your story has a solution that shows good Digital Citizenship!
4. Partner two rolls and repeats the process.

	Character	Setting	Problem
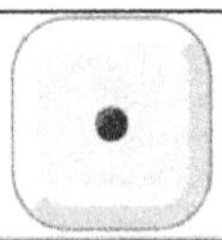	Digi	Home	Sees someone being unkind online
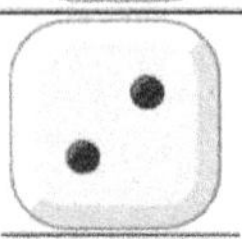	Your best friend	Home	Asks for a password to a website
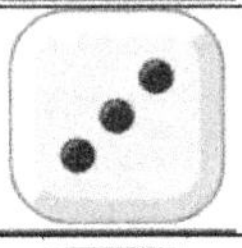	You	Playground	See someone being mean to a friend
	A parent	Home	Finds out that a child has been on inappropriate websites
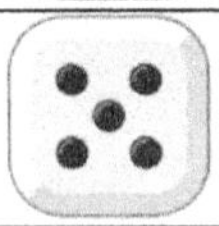	A parent	At work	Finds out their child was staying up late to play on a tablet
	A cousin	On vacation	Shows you a secret account to a website not meant for kids

DIGITAL CITIZEN ACADEMY

Grade 4 Activity

Name: ______________________

DIRECTIONS:
You have all met DIGI, Digital Citizen Academy's Model Digital Citizen. DIGI needs an assistant to help him spread the word about being a good digital citizen. ***Create a superhero to assist DIGI as he helps others become good digital citizens.***

What is your superhero's name?

What are their superpowers?

How will they help DIGI spread the word?

What do they wear?

Add words and phrases that relate to Digital Citizenship.

Be creative, use colors, and your imagination! Be prepared to share your superhero with the group. Use the space on the back page as your rough draft.

www.digitalcitizenacademy.org

DIGITAL CITIZEN
ACADEMY

Grade 7 Activity

Name: ______________________

DIRECTIONS:

Illustrate what cyberbullying looks like in terms of being repetitive, intense, hurtful, and purposefully damaging.

DIGITAL CITIZEN ACADEMY

Grade 8 Activity

Name: ______________________

DIRECTIONS:
Hold a press conference about Digital Reputation. In the space below, take notes on questions and answers.

REFERENCES

Digital Distress; Growing Up Online by Dr. Lisa Strohman and Dr. Melissa Westendorf, 2021

Hartman, Steve, "Kindness 101" CBS News (video series); ***cbsnews.ccom/feature/kindness-101-.steve-hartman/***

Website: Digital Citizen Academy; ***https://digitalcitizenacademy.org/***

Website: Dr. Lisa Strohman: Mental Health Refined;

www.drlisastrohman.com

Website: Internet Safety 101, Enough Is Enough; Making the Internet Safer for Children and Families; ***https://internetsafety101.org/***

Website: KD Resources: Creating World Changers;

https://www.kdresources.org/

Website: Webroot, Child Safety;

https://www.webroot.com/pdf/Prd_Chi_InternetUsageandDangerStats.pdf

Made in the USA
Monee, IL
12 November 2026

68350825R00059